CONTENTS

3

What is a habitat?

A place where plants and creatures live is called a **habitat**. Woods, hedgerows, estuaries and seashores are all habitats. Any living **ecosystem** contains many different types of habitat.

Butterflies can often be seen collecting nectar in hedges and fields.

Creatures and plants in these habitats live and work together, like people in a village. Each of them is important for the village to work properly. They all need each other, and if one of the villagers disappears or becomes sick then the whole village may suffer.

Many of the creatures living in one habitat feed on animals and plants from another. An owl might not be able to find enough food near the tree where it lives, and so might visit another habitat. This could be the edge of a field or pond, for example.

This bird has just caught a fish.

Very few creatures live on their own. Most live in groups. Some eat plants, some eat animals, and some eat both.

What is an ecosystem?

This tree is a habitat for many plants and animals

Ecosystems can contain several habitats where plants, animals, birds and insects work together. A forest is a large ecosystem. For example, trees and grassy banks are habitats within the forest.

Creatures (including humans) affect each other's habitats. For example, a spider might spin a web to catch its dinner some distance from where it lives.

The balance of nature

Cutting down a tree or just dropping crisps on the playground can disturb or change the balance of nature.

Some changes can cause major problems. Spraying a field with chemicals may kill lots of insects. This might mean that some birds will have to go without food.

Birds will search in various places to find food.

A school pond is an excellent place to find insects and tiny water creatures.

Finding a home

Life for plants and animals in towns and cities can be hard. Most **species** have been able to survive by changing the way they live. Many live on waste ground. Others live in or near schools.

The wall around a school is a favourite place for plants. These include moss and several types of grass. Another good place is the school pond.

Dark corners and out-of-the-way places such as store sheds or cellars also provide homes for mice, rats and other small creatures.

Changes in the seasons

Changes during the year affect the various habitats and ecosystems found around a school. During the long summer holidays many plants can grow without being disturbed. Weeds develop in the cracks in the playground and small creatures search for food.

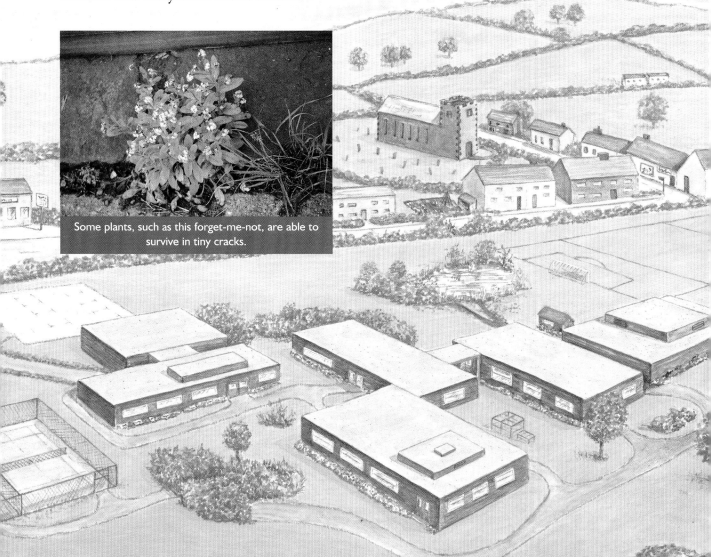

Some plants, such as this forget-me-not, are able to survive in tiny cracks.

Human environments

Schools can provide lots of food for birds and small creatures. Scraps dropped at playtime can provide food for local birds. Crows, gulls and sparrows get together just before the end of playtime, and then swoop down on leftover pieces of food once the playground is empty.

Seeds can be spread by the wind, by animals or by humans. Walking through a wild area, grass **awns** may get caught on a coat or jumper. These are then carried to another area where they may germinate.

A bluetit is making a nest in this box.

Many plants and small creatures live in the water and around the edges of ponds. Some pond creatures eat other creatures in the pond, while others eat the dead or decaying remains of plants or animals. Ponds provide an excellent food supply for newts and tadpoles.

Creating your own environment

Often schools create their own environments with a variety of habitats to encourage the spread of animals and plants. These can include wild areas or ponds.

Wild areas can be planted with flowers to attract butterflies. Small logs and large stones also make good homes for woodlice and other insects.

Nesting boxes or a simple well-supplied bird table in a quiet part of the school grounds will provide a chance to watch many different species of birds.

These birds are gathering at playtime ready to eat any leftovers after the children go in.

9

Changing times

Wild flowers can often be found on waste ground in cities and towns.

The verges of roads and motorways are a breeding ground for many creatures and plants. These include ragwort and dandelions.

Changing their ways

City creatures and plants are continually changing the way they live. If they can't find suitable trees or nesting sites, birds will look for other places to live. These include disused buildings and waste land. The robin is very good at this, and often builds nests in unusual places. Nests have been found inside the bonnet of an abandoned car and high up on a crane.

What is a food chain?

Plants and animals are all part of a **food chain**. For example, in a city a dandelion might grow in a crack; an insect feeds on the dandelion and a bird eats the insect. The bird is later caught by a fox and eaten, completing the chain. A group of food chains is called a **food web**.

A food web

In this chain, the plants are the **producers**, using the sun's energy to produce food. In turn they provide energy for insects and animals, who are the **consumers**.

Wherever there are fallen trees, **decomposers** (such as bacteria or fungi) will be at work.

Town foxes

Some creatures have changed their way of life to fit in with their new habitat. Feral foxes now live in and around cities, getting their food from dustbins and rubbish tips. Many of these live at the bottom of gardens, under compost heaps or in hedgerows. Often they can be seen or heard at dusk as they search for food.

What is a hedgerow?

Hedgerows are made up of lots of plants in a line. They need to be thick and woody to provide protection. Hedges may have shrubs, bushes and trees growing in them.

Some hedges are made from a single species of plant. This could be beech, privet or holly. Others are made from many different plants. The hawthorn is one of the most common plants used in hedgerows. It grows quickly and is very strong.

Over 200 species of plants can be found along hedgerows, including climbers (such as honeysuckle), goose grass, cow parsley and stinging nettles. The plants provide shelter and food for many of the insects and other creatures that live in and under the hedge.

Hedges often become overgrown with a mixture of flowers and plants.

Plants and animals

The types of plants and animals in a hedgerow depend on where it is. Farm hedgerows may contain lots of brambles and stinging nettles, while the hedge of a cottage may include a mixture of wild and garden flowers.

Insects and plants that live along the hedgerow have adapted themselves to their environment and depend upon one another.

Small **mammals** that live in the hedgerow use the thick cover for protection. Other creatures feed on the plants.

How did they start?

Hedges were first used to keep some animals in and some out. They also marked the boundaries of people's land.

This hedgehog is looking for food under a hedgerow.

This hedge brown butterfly has settled on some brambles.

All around the country hedges have been cut down to make larger fields.

The danger of disappearing hedgerows

Many of our hedgerows are disappearing because farmers need more space and larger fields. A few years ago lots of hedges were destroyed to make fields bigger. A good hedge can provide a home for birds, small creatures, spiders and insects, including beetles and caterpillars. Blackbirds, thrushes and chaffinches use hedgerows as nesting sites. These insects and birds help the farmer by eating the **pests** which feed on the farmer's crops.

In the countryside hedgerows act as 'wildlife corridors': long narrow wild places that criss-cross the countryside, joining one habitat with another.

Looking after a hedge

Most farmers cut their hedges every year to keep them dense.

Hedge-laying takes place in the winter months. The hedge-layer uses a large curved knife, an axe, thick gloves and a **billhook**.

A Somerset hedge-layer at work

All of the old dead wood is taken away. The branches are then woven around thick posts. A strong hedge is produced, which can provide homes for many different plants and creatures.

Hedgerows stop the land from wearing away and provide many rich and living habitats.

This hedge has been laid.

This is a long-eared bat.

Creatures of the night

Many of the creatures that live in and under the hedgerow are **nocturnal**. This means that they come out at night and sleep

Greenfinches enjoy eating hawthorn berries along the hedgerow.

during the day. Creatures that live along the hedgerow during the day, such as butterflies or greenfinches, disappear at night and are replaced by moths and bats.

Voles and hedgehogs

Voles live in holes beneath the hedgerow and feed on fungi, nuts and berries. Hedgehogs (so called because they spend a lot of time along hedgerows) grub about looking for insects. They often **hibernate** in the fallen leaves.

Mice

Woodmice and harvest mice live in hedgerows that border fields. Harvest mice eat the grain from the fields, and woodmice make their homes under the **leaf litter**.

A vole

Other creatures

Foxes regularly visit hedgerows looking for food.

Rabbits living along the hedgerow may damage it by eating the young shoots or removing soil and roots as they dig their burrows.

There are often signs during the day that animals were around the previous night. These can include fresh droppings, half-eaten fruits, wisps of hair, footprints in the mud and pathways through the undergrowth.

All the creatures living in and around the hedgerow depend on it for their survival.

What lives in a wood?

A large number of plants, animals, birds and insects live in woodlands. They are rich in food and can provide shelter and safety from predators.

Woodland trees will often be host to lots of small animals, plants and fungi.

Deciduous woods

In woods made up of **deciduous** trees, the falling leaves in autumn provide an excellent home for a wide variety of insects. Woodlice, which like damp, dark conditions, can be found under rotting leaves and logs. Other small creatures that enjoy this part of the wood include millipedes and stag beetles.

Many of the plants on the woodland floor have large leaves which collect as much sunlight as possible.

Decomposers such as fungi are continually at work on the woodland floor breaking down the dead plants and animals.

The soil around the base of a tree can be rich in minerals and provide a home for woodland plants. Some of these, such as wild daffodils and bluebells, have developed ways of storing food in bulbs during the winter. They can later release this food when they need it in the spring.

Although there is less sunshine in the winter, the trees let through more light because they are bare of leaves. This helps the plants which grow on the woodland floor.

An old oak tree

This Scottish wood is a dark place in summer as the leaves on the trees block out the light.

Coniferous woods

In **coniferous** woods there are fewer plants around the base of the trees. This is because there is less light during the year and less nourishment in the soil. Insects are still plentiful in these woods.

Adapting to woodland

Many of the creatures that live in woods have adapted to their environment. Woodpeckers have developed special beaks that can drill into trees in search of beetle larvae and other wood-boring insects. Woodlice have developed tiny gills which help them survive under stones and logs. Crossbills have cross-over beaks to take the seeds out of cones.

A woodpecker

19

A natural wood in Sussex

Managed woodland

Many of the woods in Britain have been **managed** in some way for thousands of years. Wood from them was used for building homes. The trees in the New Forest were used to build huge wooden warships. In Somerset, willow trees have been grown for

A managed wood

many years to provide wood for basket-making and charcoal.

What is woodland life like?

Life in these managed woods can be very different from that in natural woodlands which have been allowed to grow with little or no interference from humans.

20

In some managed forests trees are grown for their wood. These trees are often of a single type. Their tops can form a thick canopy which cuts out light from the forest floor. However, many plants are still able to live under them, and there are squirrels and even wildcats.

In some places trees are 'coppiced'. Wood is taken from a tree without killing it. The tree is cut down almost to the bottom and then allowed to grow again. In this way the tree can continue to produce wood for a long time.

A managed wood

Over the centuries, a great deal of British forest has been cut down without being replanted, and today Britain is one of the least wooded countries in Europe.

What happens to trees after they are cut down?

After the trees from managed woods have been chopped down using a chainsaw or special machine, they are taken to a sawmill. Here the bark is stripped and the wood is turned into timber. Every part of the tree is used: even the bark is cut up and bagged to be used on gardens.

How old is it?

You can tell the age of a tree by counting the rings. As the tree grows it produces a new layer of young wood behind the bark. When the tree is cut down, the rings can be counted to find out how old it is. Another way of working out the age of a conifer is to count the number of rings of branches it has grown over the years.

Here you can see the growth rings on a tree.

Dead tree stumps

After a tree has fallen, it is often not long before new shoots begin to appear around the base. These provide cover for the many creatures that live on the woodland floor.

There are often lots of old tree stumps and logs on the forest floor where trees have been cut down or have fallen over.

Many of these stumps and logs provide special habitats where fungi, plants, bacteria and insects can work together to help leaves and animal matter to decompose. These creatures work day and night, growing or feeding on the dead material that they are helping to break down.

They like to build their setts in secret places away from humans, although over the past few years more and more have come into villages and towns in search of food.

Birds have also adapted to woodland life by developing short wings which help them move between the trees. Blackcaps enjoy eating the young saplings that grow up around tree stumps.

Somewhere to live

Dead branches and rotting logs also provide homes and food for a number of small animals and birds. Half-eaten pine cones show that squirrels have been in the area. In some woods, mounds of earth around a fallen log or a series of holes in the bank may indicate a badger's **sett**.

Badgers and birds

Badgers are an important part of the food web. They will eat berries, worms and even dead deer.

These badgers are coming out of their sett.

As the human population increases we need to produce more food. As a result, farms have become much larger. Hedges have been removed so that bigger crops can be grown, and huge machines are used in the fields.

Many lowland farms keep dairy cows; others grow crops, and some do both.

This field is full of ripe crops.

Some lowland farms have large herds of dairy cows.

On **dairy farms**, using special machines, one person can now milk over 100 cows in just two hours. On **arable farms** a combine harvester can harvest a field of wheat in just a few hours.

These changes have meant that many of the hedgerows are no longer needed to separate fields or act as windbreaks. However, not all of the hedgerow plants and creatures have disappeared: some of them have managed to change the way they live.

Life on a cowpat

Some flies will have a meal from this cowpat.

There are some surprising habitats on farms. A field of cows will usually contain lots of cowpats. These have their own tiny communities of micro-organisms and creatures, which help break down the cowpat, in turn helping the grass to grow.

Life in a field

Harvest mice can be found in fields, making nests by weaving together strands of living grass. Field voles can be found on the edges of fields, together with shrews and larger creatures, including deer.

Today many farmers are replanting hedges and setting aside areas where plants and creatures can live without human interference.

These young harvest mice are playing in a field of wheat.

These upland farm buildings are well built and strong.

Upland farms

Upland farms are very different from those lower down.

The barns and farm buildings are often much stronger as the weather can be very severe.

High uplands include moorlands, where the ground is covered in moss, bracken or rough grass. Many upland farmers breed sheep or cattle which are used for meat. Moorland farming can be hard work because much of the ground may be rocky or boggy and in winter there may be heavy falls of snow.

Farm animals

Many of the sheep and cattle that live in the hills and on moors are special breeds that are used to the cold and often windy weather.

They have special teeth to deal with the tough grass, and thick coats for the rough weather.

Hill farming can be tough for the farmer, who knows that spells of bad weather can mean the death of sheep or cattle.

This highland cow is protected against the weather by its long thick coat.

Cattle and sheep may be left during the summer to graze in the higher fields and then brought down for the winter.

What kinds of plants live higher up?

Plants found on hillsides and moors have adapted themselves to life on high ground. Some plants do not open their flowers until they are certain that the snow has gone. Many grow close to the ground to protect themselves from the weather.

Heather has roots that spread out sideways rather than growing downwards. This means that it is able to collect rainwater more easily. It also has very small leaves to prevent it from losing too much water.

What are meadowlands?

Meadowlands can be used for grazing animals or making hay, but there are few real hay meadows left today. This is because many cattle are now fed on silage, which is young green grass that has been cut and stored.

Meadows can be found on low or high ground. They often contain a great deal of plant and insect life, including poppies, bees and butterflies. Unless they are used

for grazing animals, they need to be cut once a year, otherwise trees will start to grow and turn the meadow into a wood. Meadow plants also include clover and various types of wild grasses, which grow quickly after they have been cut.

What kind of life will you find in a meadow?

Animals and plants can live in meadows with little human contact. Near the ground are the snails and frogs, which like the damp, dark conditions. Higher up are the caterpillars and grasshoppers, and at the highest level are bees and butterflies, which gather nectar from the flowers.

Daytime visitors include swallowtails, peacock butterflies and meadow pipits. At night, small mammals, including hedgehogs, visit the meadow.

Life also goes on underneath the meadow. Small mounds of earth are a sign of moles. Other creatures such as earthworms live in the soil, and mole crickets may be at work below the surface of the meadow eating the roots of the plants.

Some farmers use their fields in rotation so that some are left as meadows for grazing while others are cultivated for crops.

A meadow pipit

Over the past few years many schools have made their own meadows where they can study plants and animals.

A school wildlife area

Spraying fields

Farmers sometimes spray their fields with chemicals to get rid of pests or make the crops grow.

This aeroplane is flying low so that the chemicals will fall on the crops.

Chemicals are also used to keep down the weeds. But some may kill creatures that are important in the food chain. Wiping out all pests can be dangerous, as their predators will have nothing to eat and will die. This may cause problems further up the food chain.

Destroying food chains can affect the more complicated food webs. Animals and plants from one chain may feed from another. By destroying one chain it is possible to harm many more.

Crops can also be sprayed from the air, but changes in the wind can cause chemicals to land on the wrong plants. It is important for the farmer to check the direction of the wind before spraying.

These chemicals are being sprayed on the land using a tractor.

As an alternative to spraying, the farmer may introduce a new predator to control pests. But there is then a danger that the new species may itself go out of control.

Chemicals sprayed on the plants may soak into the land and eventually get into the water supply, or be washed into rivers and streams.

It is important that farmers should be very careful how they spray their crops. 'Organic' farmers use only natural substances (such as manure) to help their crops grow.

Chemicals drain off the fields and into streams and rivers.

This farmer uses only natural fertilisers to help his plants grow.

This is the Wash estuary in Norfolk.

Estuaries

Estuaries provide some of the richest habitats in the world. Here fresh water from a river meets salt water brought into the estuary by the tide. Mud, pebbles, grit and dead plants brought down by the river mix with the salt water to create a special kind of ecosystem which supports a great variety of life. The mud and silt that are produced make an excellent breeding ground for small fish, worms and creatures, such as cockles, which suck up the mud.

Food for all

The large supply of fish and small creatures provides food for large numbers of birds. Many of these are visitors which may stay only a few months before flying to other parts of the world such as Asia and Africa. These include ducks and geese. Some birds, such as dunlins, travel long distances south to breed and use estuaries as places where they can stop and feed. Another regular visitor is the black and white, orange-beaked oystercatcher.

This duck has a specially shaped bill to help it feed.

Birds that live in estuaries often have specially shaped bills. For example, curlews and godwits have very long bills which they use to dig deep into the mud for lugworms. Some wading birds have long legs to keep them above the water.

Sea lavender is an estuary plant.

Much of what goes on in an estuary is controlled by the tides. Birds that cannot swim have to leave the mud flats when the tide comes in and return later.

Estuaries also have lots of plants, such as sea lavender and yellow-horned poppies, which are specially adapted to the changing amounts of salt in the water. Many also have taproots which go deep into the mud.

This oystercatcher uses its long beak to break open shells.

Large numbers of birds collect together close to the water to feed.

A food chain

Food for all

Every day the tide brings an enormous amount of food into an estuary. This includes shellfish, worms, crabs and shrimps. Millions of tiny **plankton** also arrive.

These plankton are eaten by the shellfish and worms. These are then eaten by the wading birds, completing the food chain.

Shelducks collect tiny snails by sieving water with their specially shaped beaks. Oystercatchers open mussels and oysters with their strong beaks.

Who visits the estuary?

Scientists and researchers spend a lot of time on the mud flats taking samples and examining the creatures in them. There are thousands of shellfish and worms hiding in every square metre of mud that gets covered as the tide comes in.

Another human visitor to the estuary is the bait digger. Bait diggers dig in the mud collecting lugworms to use as bait for fishing.

What kinds of problems are there?

Although estuaries are excellent places for birds to feed, they can also be dangerous for birds. Oil spills from tankers, or pollution from factories further up the river, can poison the water and kill many birds. Sometimes, in bad weather, food supplies may be used up or birds may die from the cold.

These children are carrying out research in the mud of the estuary.

This bait digger is digging deep into the mud looking for worms.

Coastlines provide a variety of habitats, from rocky shores to sandy beaches. On the beach some of these are covered by the sea for twelve hours a day, while others closer to the sea are under water for most of the time.

Puffins make their nests high on the cliffs.

A hard life

Life along the seashore is hard. Plants near the sea are sprayed with sea water, and those high up on the cliffs may have to find cracks in the rocks where they can take root. Plants on top of cliffs have learnt to survive by sending their roots out along the ground, or by growing close to the ground to avoid the wind.

Cliffs provide safe homes for millions of seabirds, with herring gulls and puffins nesting on cliff ledges.

Many of these birds spend time searching for food in the sea, while along the shoreline smaller birds such as the rock pipit and oystercatcher look for food amongst the stones, seaweed and rock pools.

Life in a rock pool

Rock pools

Rock pools provide some of the richest habitats along the beach, containing many different plants and creatures. These include sea anemones (which are not plants, but animals). Hermit crabs can also be found living in old sea shells. They have a soft body which needs protecting. When the crab get bigger it moves on to a larger shell.

This hermit crab has made its home in a shell.

Sand casts

Many of the creatures that live on sandy beaches cannot be seen, because they bury themselves beneath the sand for protection. However, there are often clues in the sand that creatures are below, such as coils of sand or tiny dips. These may be produced when the creature is feeding or breathing. Lugworms leave coils of sand as they bury themselves, and a tiny dip in the sand may hide a concealed crab.

One of the toughest places on the beach is the part which is covered by sea water twice every 24 hours.

Shoreline plants

Many of the plants here have adapted themselves to the daily tides. Seaweeds have tough skins and are able to bend easily as the water flows over them. Some types have special air bladders which allow them to float. Others have 'holdfasts'. These are not roots, but special points which fix on to stones or rocks. Seaweeds also produce a slimy liquid which stops them from drying out in the air.

When the tide comes in it can bring lots of new visitors to the beach.

Life along the beach

These fishing boats are pulled up along the shore at Hastings in Sussex.

The creatures and plants along the upper shore may only be covered for a few hours, while those further down may spend half the day under water. Many of these creatures provide food for seabirds living close to the beach.

Along the top of the beach you may find the bones of small creatures such as cuttlefish, broken pieces of seaweed, and shells washed up by the sea. This part of the beach is regularly visited by seabirds looking for food.

Humans play an important part in the food chain. Fishing boats pulled up along the shoreline provide plenty of food for seagulls and other birds. Seabirds have learned that these boats provide an easy meal, and will swarm around them as they come ashore.

This seagull has caught a starfish.

Over the years, creatures and plants have had to adapt, changing the way they live – and even their colour.

One of the best examples of this is the peppered moth. About 200 years ago, when the trees in cities were covered in dirt and grime, the dark-coloured moths found it much easier to survive than the silver-coloured moths. This was because they could not be seen so easily by predators. So more dark moths survived and passed on their characteristics to their offspring. However, in the past few years more silver moths have appeared. This may be because our cities are starting to become cleaner.

This is the Channel Tunnel being built.

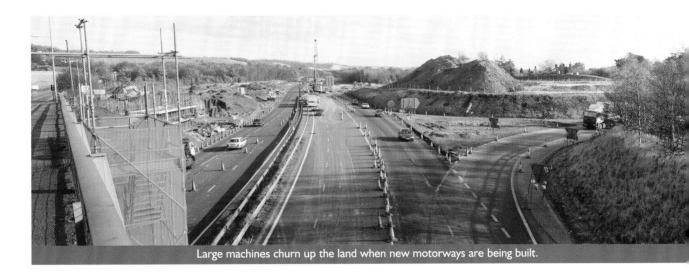

Large machines churn up the land when new motorways are being built.

Although many animals and plants are able to change the way they live, it is still important for us to protect their natural habitats.

When new buildings are put up, plants and small creatures are often moved before the work starts and returned when it is finished.

It is important to replace trees that have been chopped down and to plant new hedgerows. Many farmers are now using fewer chemicals on their crops.

In towns and cities, more and more wildlife areas are being created.

We can all help by putting up bird tables and nesting boxes, respecting the countryside and joining groups that are setting up wildlife reserves and conservation areas.

The world is like a web which is connected together. It is important that the strands of this web should not be destroyed.

We have only looked at a few contrasting habitats. There are plenty more for you to discover!

arable farm. A farm where crops such as wheat and barley, or root crops like sugar beet, are grown.

awns. Bristly spikes produced by grasses such as wall barley that fix themselves to animal fur or human clothing as it brushes past them.

billhook. A special tool used by a hedge-layer. Different types of billhooks are used in different parts of the country. The hedge-layer cuts off the side shoots of the hedge with the billhook.

coniferous trees. Trees that keep their leaves throughout the year.

consumers. Animals which eat plants ('primary consumers'), or which eat other animals ('secondary consumers').

dairy farm. A farm where cows are raised and kept for their milk. The cows are milked regularly. The milk is taken to a dairy where it is bottled or put into cartons. A popular dairy cow is the black and white Friesian.

deciduous trees. Trees that lose their leaves every year.

decomposers. Organisms whose job is to break down the remains of plants and animals. They include bacteria, some insects and fungi.

ecosystem. A community of plants and animals. It is often made up of several habitats. It includes the surface, the soil and rocks under the ground, and the air above it.

food chain. A chain of creatures or plants that normally feed on one another.

food web. Several food chains linked together.

habitat. A place where plants and animals live.

hibernation. Sleeping through the winter months.

leaf litter. Matter often found on the woodland floor, made up mainly of fallen leaves.

mammal. A warm-blooded animal whose female feeds her young with milk. It has a backbone, and usually has hair.

managed forests. Forests which have been planted by people. The trees are then cared for.

meadowlands. Grasslands, often used to make hay or for animals to graze on.

nocturnal creatures. Creatures that come out at night.

pests. Creatures that damage crops and plants.

plankton. Tiny creatures that live in the sea.

producers. Green plants which make their food simply using sunlight, carbon dioxide from the air and minerals from the soil.

sett. The underground home of a badger.

species. A group of plants or creatures of a single type.

Useful books

Steve Pollock. *Ecology*.
Dorling Kindersley, 1993.

Richard Spurgeon. *Ecology*. Usborne,
1988.

Tess Lemmon. *Woodlands*. BBC,
1992.

Barbara Taylor. *Shoreline*.
Dorling Kindersley, 1993.

Michael Clark. *Tracker Series:
Mammals*. Hamlyn, 1991.

Andrew Cleave. *Tracker Series:
Seashores*. Hamlyn, 1991.

Paul Sterry. *Tracker Series: Butterflies
and Insects*. Hamlyn, 1991.

Rob Hume. *Tracker Series: Birds*.
Hamlyn, 1991.

Gwen Allen, Joan Denslow.
Cluebooks Series. Oxford, 1997.

Various. *Jump Ecology Series*.
Two-Can Publishing, 1995.

Terry Jennings. *People or Wildlife*.
A&C Black, 1996.

Gillian Symons, Prue Poulton.
The Urban Environment.
A&C Black, 1996.

Lynne Pratchett. *Trees for Tomorrow*.
A&C Black, 1996.

Useful addresses

Junior RSPB
The Lodge
Sandy
Beds SG19 2DL

Woodland Trust
Autumn Park
Dysart Road
Grantham
Lincs NG31 6LL

National Federation of Badger Groups
15 Cloisters Business Centre
8 Battersea Park Road
London SW8 4BG

Elm Farm Research Centre
(Organic Farming)
Hamstead Marshall
nr Newbury
Berks RG20 0HR

Countryside Education Trust
John Montagu Building
Beaulieu
Brockenhurst
Hants SO42 7ZN

WATCH
The Green
Witham Park
Waterside South
Lincoln LN5 7JR

Somerset Farming and Wildlife
Advisory Group
County Hall
Taunton TA1 4DY

Acknowledgements

We acknowledge the copyright of the following people and institutions, and thank them for their kind permission to reproduce their pictures:

Byrne Cragie Photos
pp. 10a, 41a

BBC Natural History Unit Picture Library
pp. 4b, 13b, 16b, 32a, 38a, 39b

Elm Farm Research Centre
p. 31b

Eurotunnel
p. 40b

Forestry Commission
p. 21a

Peter Holden
pp. 4a, 5a, 6a, 7a, 8a, 12b, 14a, 15a, 15b, 19a, 20a, 21b, 24a,
24b, 25a, 26a, 26b, 34a, 35a, 36a

Natural History Museum
p. 40a

Oxford Scientific Films
pp. 10b, 12a, 13a, 16a, 17a, 18a, 18b, 20b, 22a, 22b, 23a, 23b,
25b, 27a, 29a, 33a, 37a, 37b, 39a

RSPB Images
pp. 5b, 9a, 10b, 32b, 33b, 35b

Still Pictures
pp. 30a, 30b, 31a

Published by Channel Four Learning Limited
Castle House
75–76 Wells Street
London WIP 3RA

Designed by Andrew Barron & Collis Clements Associates
Original illustrations by Jane Bennett
Printed at the Bath Press

ISBN 1862152705

**This book is to be returned on or before
the last date stamped below.**

Harvey Woolfe and Evans Woolfe.